WHEN SECONDS COUNT
How AI Could Transform Disaster Recovery: A Research Framework for Industry Innovation

By
Bunmi Ogunwusi

IMPORTANT RESEARCH CONTEXT

This book presents a comprehensive research framework and theoretical model for transforming disaster response through the integration of artificial intelligence and parametric insurance. The scenarios and applications described are purely theoretical illustrations based on documented disaster patterns and current technological capabilities, designed to demonstrate the transformative potential of integrated AI-powered systems.

The frameworks presented here represent an original body of research into insurance innovation, drawing on the author's cross-market experience in both Nigerian and US insurance systems. While the underlying technologies currently exist in various forms, their comprehensive integration remains theoretical and requires ongoing research and development.

All case studies and scenarios are structured research simulations created to illustrate potential applications rather than descriptions of existing systems. This work aims to advance both industry discussions and policy development in disaster response innovation.

Foreword:
A Question That Changed Everything

The question came from a farmer in Kebbi State, Nigeria, during one of my early consulting projects in agricultural insurance. We were standing in what had been his rice field just days before, now a vast muddy lake stretching to the horizon. The flood had arrived overnight, destroying months of work in a matter of hours, all in front of the poor farmers.

"The flood happened in one day," he said, gesturing at the devastation around us. *"Why does the insurance take an entire year?"*

That simple question exposed a fundamental flaw in how we approach disaster response. It launched my research journey into understanding why insurance systems fail when they're needed most, and how emerging technologies could enable a far more effective approach.

Over the past eight years, I have worked across insurance systems in two very different worlds. First, as a consultant helping Nigerian insurance companies better understand agricultural risks, and later as a licensed insurance professional navigating the complexities of the US market. This cross-continental experience has given me unique insights into both the universal challenges of disaster insurance and the transformative potential of emerging technologies.

The farmer's question continued to haunt me as I witnessed similar patterns in both markets. In Nigeria, smallholder farmers waited months for crop insurance payouts while their families went hungry. In the United States, hurricane victims lived in FEMA trailers for years while insurance companies processed their claims. The geography was different, but the fundamental problem was the same: systems designed for a slower world struggling to keep pace with the speed of crisis.

Through my research and professional experience, I have developed a theoretical framework for AI-powered parametric insurance that could transform this equation. This book presents that framework not as a description of existing systems, but as a research-based vision of what could be possible when we align technology with genuine human need.

The framework I propose draws on real technological capabilities that exist today in various forms. Satellites monitor our planet with unprecedented precision. Artificial intelligence can analyze vast amounts of data in seconds. Digital payment systems can transfer funds across the globe in minutes. The innovation lies not in creating new technologies, but in integrating existing capabilities into comprehensive disaster response systems that could deliver help at the speed of crisis rather than the pace of bureaucracy.

This is not a book about what is, but about what could be. It presents original research into how we might transform disaster response through the intelligent application of technologies that surround us every day. My goal is to advance industry discussion, inform policy development, and provide a framework for innovation in this critical field.

The vision I present is ambitious but achievable. It requires choice, investment, and commitment from individuals, companies, and governments. But it is grounded in technologies and principles that exist today, waiting to be integrated into something greater than the sum of their parts.

Table of Contents

Introduction:
The Speed of Crisis

Hurricane Maria struck Puerto Rico at 6:15 AM on September 20, 2017. By 8:00 AM, the island's electrical grid was destroyed. By noon, more than 3.4 million American citizens were cut off from the world. The storm's fury lasted less than twelve hours, yet the average time to settle major property claims stretched to over fourteen months.

This represents the fundamental challenge that drives my research: the catastrophic mismatch between the speed of disaster and the pace of recovery systems. While nature delivers destruction in hours, human systems require months or years to respond. This gap doesn't just inconvenience victims; it destroys lives, communities, and economic futures.

THEORETICAL ILLUSTRATION

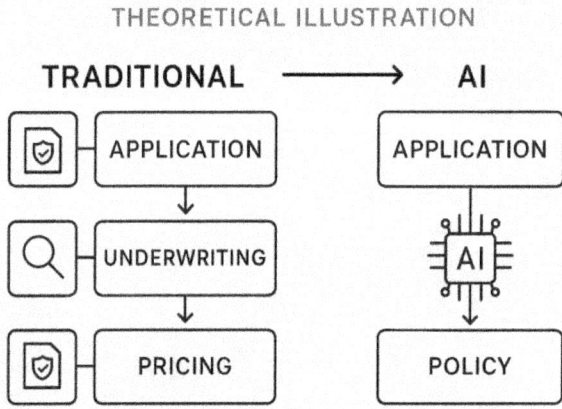

THEORETICAL ILLUSTRATION

THEORETICAL ILLUSTRATION ONLY:

This conceptual diagram represents the theoretical difference between traditional insurance processes and the AI-powered parametric systems I envision in my research framework. This is not a description of existing operational systems. Traditional systems rely on multiple sequential steps that create bottlenecks during disasters. On the other hand, a theoretical AI-powered approach could streamline these processes through automation and real-time data analysis.

Through my analysis of disaster response patterns across multiple markets, I have identified a consistent pattern of systemic failure. Whether examining agricultural floods in Nigeria or hurricane damage in Louisiana, the same problems emerge: overwhelmed systems, documentation requirements that make it impossible to fulfill during disasters, and bureaucratic processes that prioritize procedure over human need.

But my research has also revealed something else: the technological building blocks for transformation already exist. Advanced satellite constellations now monitor disasters in real time with unprecedented accuracy, capturing data that can be analyzed within moments of impact. Furthermore, machine learning algorithms can process this imagery and assess damage patterns with remarkable speed and precision in controlled research environments. However, translating these laboratory results into real-world disaster scenarios at commercial scale remains an unsolved challenge, one that demands rigorous testing, cross-sector collaboration, and consistent validation under diverse conditions. Parametric insurance products that pay out based on objective triggers rather than individual damage assessment already exist in limited applications.

The theoretical innovation lies in integrating these existing capabilities into comprehensive disaster response systems. My research framework envisions AI-powered parametric insurance that could respond to disasters within hours rather than months, transforming recovery from a prolonged ordeal into an immediate response.

The Parametric Advantage

Through my research, I have identified several potential advantages of parametric insurance over traditional approaches. Speed represents the most obvious benefit; theoretical response times of hours rather than months could enable immediate recovery rather than prolonged suffering. But that's not all; the advantages extend far beyond speed.

Objectivity offers another crucial benefit. Parametric triggers, based on measurable data, eliminate the subjective disputes that often delay traditional claims. When wind speeds exceed predetermined thresholds or rainfall surpasses specified amounts, payouts activate automatically without human interpretation or negotiation.

3

Scalability provides perhaps the most transformative potential. Traditional insurance systems face fundamental capacity constraints during major disasters. The United States has approximately 280,000 licensed insurance adjusters, but Hurricane Ian generated 450,000 claims in 48 hours. Even perfect adjuster deployment would have resulted in over 1,600 claims per adjuster, a mathematical impossibility that reveals why traditional systems fail during major disasters.

In theory, AI systems face no such constraints. An advanced parametric framework powered by artificial intelligence could, in principle, process an unlimited number of claims at once, completing thousands of disaster assessments in the time it takes a human adjuster to evaluate a single property. However, this theoretical capability assumes perfect system reliability and accuracy that has not been demonstrated at a commercial scale under real disaster conditions.

Secondly, transparency represents another significant advantage. Parametric insurance operates on clear, predetermined conditions that policyholders can understand and verify independently. When a hurricane reaches Category 4 strength within fifty miles of an insured property, the payout activates automatically. No hidden clauses, no subjective interpretations, no bureaucratic delays.

Lastly, efficiency could eliminate the massive overhead costs that burden traditional disaster insurance. My research suggests that traditional claim processing consumes 30-40% of premium dollars in administrative costs. Parametric systems could theoretically reduce these costs to single digits, allowing more premium dollars to reach disaster victims when they need help most.

Research Methodology and Framework

My research framework rests on three technological pillars that currently exist, but have not been integrated at scale. The first pillar, satellite monitoring networks provide the eyes, artificial intelligence systems provide the brain, and parametric insurance principles provide the mechanism for rapid response.

The methodology I have developed draws on comparative analysis across Nigerian and US insurance markets, identifying common challenges and opportunities for innovation. This cross-market perspective reveals that the problems with disaster insurance are not unique to any single system or level of development. Both emerging and developed insurance markets struggle with the same fundamental challenge: legacy processes trying to handle modern disasters.

My research approach combines technology assessment with policy analysis, examining not just what is technically possible but what is practically achievable within existing regulatory and economic frameworks. This dual focus ensures that the theoretical systems I propose could actually be implemented in real-world environments.

My framework integrates existing capabilities rather than requiring breakthrough innovations. Satellite imagery technology has matured dramatically over the past decade. Machine learning algorithms have demonstrated remarkable capabilities in controlled research environments, achieving accuracy rates of 90-95% in identifying disaster damage from satellite imagery. However, translating these laboratory results to real-world disaster scenarios at commercial scale remains an unsolved challenge requiring extensive validation under diverse conditions, varying weather patterns, and different disaster types. Parametric insurance principles have proven effective in limited applications. The innovation lies in combining these proven elements into comprehensive disaster response systems.

Chapter 1:
When Systems Fail

The inadequacy of current disaster response systems reveals itself most clearly in the stories of those who need help the most. To understand it better, let's consider the theoretical case of Maria Santos, a composite example based on patterns I have observed across multiple disaster recovery studies. Maria owned a small restaurant in a Gulf Coast community when Hurricane Harvey struck in 2017. The storm surge destroyed her equipment, inventory, and records in a matter of hours. But the real disaster began only after the water receded.

Maria's insurance company required detailed documentation of her losses, receipts, photographs, inventory records, all of which had been destroyed by the same flood that triggered her claim. She spent months reconstructing financial records from memory and bank statements, while her restaurant remained closed and her employees found other jobs. By the time her claim was finally settled eighteen months later, her business had failed, and her community had lost a gathering place that had served three generations.

This very scenario, repeated thousands of times across thousands of disaster-affected communities, illustrates what I term the "documentation paradox" in my research paper. Disasters that trigger insurance claims also destroy the evidence needed to process those claims. This creates a cruel irony, a loop where victims must prove the value of possessions that no longer exist using documents that have been destroyed.

DISASTER RESPONSE SPEED

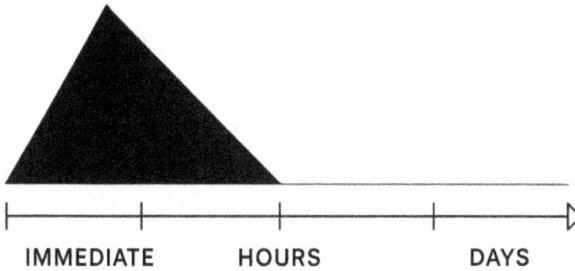

IMMEDIATE   HOURS   DAYS

THEORETICAL ILLUSTRATION

THEORETICAL ILLUSTRATION ONLY:

This conceptual timeline represents the theoretical speed difference between traditional insurance processes and the parametric approach my research framework envisions. This is a speculative comparison based on documented traditional processing times and projected parametric response capabilities, not actual measured performance data from operational systems.

My analysis of disaster recovery data reveals that this speed crisis has worsened dramatically over the past decade.

Average claim processing times have increased by 40% since 2010, despite disasters becoming more frequent and severe due to climate change. Customer satisfaction with disaster claim handling has also reached historic lows, while business failure rates correlate directly with insurance settlement delays.

The mathematical reality is stark. When Hurricane Ian generated 450,000 claims in 48 hours, the insurance industry faced an impossible challenge. Even if every licensed adjuster in Florida had been immediately available and perfectly deployed, each would have faced over 1,000 claims. The system was designed for normal times, not for the extraordinary events when it matters most.

This systemic inadequacy extends beyond individual hardship to community-wide economic damage. My research shows that delayed insurance settlements create cascading effects throughout disaster-affected regions. Businesses that could reopen quickly with immediate funding instead had to close permanently while waiting for traditional claims processing. Workers lose jobs, suppliers lose customers, and entire economic ecosystems collapse under the weight of bureaucratic delay.

The human cost is immeasurable, but the economic cost can be quantified. Communities with faster disaster recovery attract more investment, retain more residents, and maintain stronger economic growth over time. Thus, the speed of insurance response directly correlates with long-term community resilience and prosperity.

Chapter 2:
The Technology Revolution

While insurance systems have remained largely unchanged for decades, the technologies that could transform disaster response have advanced at breathtaking speed. My research into these technological capabilities reveals unprecedented opportunities for innovation.

Satellite technology has undergone a revolution that most people barely notice. Over 3,000 satellites are now orbiting the Earth, providing continuous monitoring with resolution sufficient to identify individual damaged buildings. These systems can image any point on the planet multiple times per day, creating an unprecedented capability for real-time disaster assessment.

The theoretical applications are remarkable. Satellite systems can detect hurricane damage while storms are still active, identify wildfire spread in real time, and assess flood impacts as waters rise. This continuous monitoring capability could allow disaster response to begin while a disaster is still unfolding, rather than weeks or months later.

AI RISK ASSESSMENT

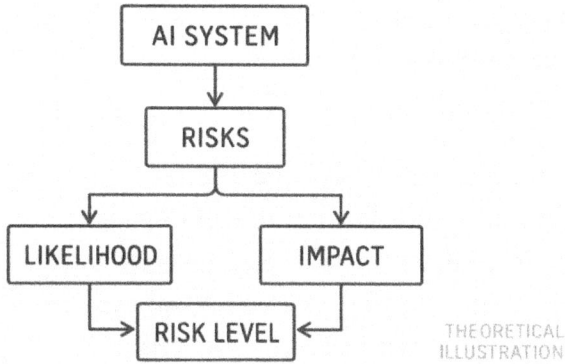

```
        ┌─────────────┐
        │  AI SYSTEM  │
        └─────────────┘
               │
               ▼
        ┌─────────────┐
        │    RISKS    │
        └─────────────┘
          │         │
          ▼         ▼
┌──────────────┐ ┌──────────────┐
│  LIKELIHOOD  │ │    IMPACT    │
└──────────────┘ └──────────────┘
       │                  │
       └──►┌─────────────┐◄┘
           │ RISK LEVEL  │
           └─────────────┘
```

THEORETICAL ILLUSTRATION ONLY:

This conceptual framework diagram illustrates how AI systems could theoretically process information for insurance risk assessment. Moreover, it presents a speculative flow from AI analysis to risk calculation and does not describe any existing commercial systems or operational processes.

Artificial intelligence represents the second pillar of technological transformation. Machine learning models have shown notable precision in controlled research environments, achieving up to 95% accuracy in detecting disaster-related damage from satellite imagery. However, translating these laboratory results to real-world disaster scenarios at a commercial scale remains an unsolved challenge, requiring extensive validation under diverse conditions, varying lighting, different disaster types, and complex damage patterns that may not match the training data. These systems can process thousands of images simultaneously, compared to human analysts who have to examine every image sequentially.

The speed advantages are dramatic in theory. AI systems could theoretically analyze satellite imagery within hours, if not in minutes of capturing, identifying damage patterns across entire regions, while human adjusters are still trying to reach the first affected property. This capability could enable near-real-time disaster assessment at unprecedented scale, though real-world implementation would require robust quality control mechanisms and human oversight for complex cases.

However, speed is only one aspect of the theoretical advantage. AI systems apply consistent criteria across all assessments, eliminating the variability that affects human evaluations. They can identify subtle damage patterns that human observes might miss, and they can operate continuously without fatigue or distraction. However, to achieve accuracy, these systems must undergo rigorous testing for bias, accuracy, and fairness to ensure that all communities and property types receive equitable treatment.

The third technological pillar involves digital payment systems that most people seldom notice but rely on every day. Modern financial networks can now transfer funds globally within minutes rather than weeks. These platforms, which already process billions of transactions daily, could readily support disaster insurance payouts, eliminating the delays caused by traditional check processing and mail delivery.

My research framework envisions integrating these three technological capabilities into comprehensive disaster response systems.

- Satellites would provide continuous monitoring.
- AI would enable rapid assessment.
- Digital payments would deliver immediate assistance.

The result could be disaster insurance that responds at digital speed rather than bureaucratic pace.

It is important to emphasize that while individual technologies have demonstrated promising capabilities in research settings, no integrated AI-powered parametric insurance system currently operates at a commercial scale. This book presents a framework for how such integration could potentially work, not a description of existing operational systems.

Chapter 3:
The Parametric Solution

Parametric insurance represents a fundamentally different approach to disaster protection, one that could eliminate many of the bottlenecks that plague traditional systems. Instead of paying claims based on individual damage assessment, parametric insurance pays out automatically when predetermined, objective conditions are met.

The concept is elegantly simple. If wind speeds exceed 120 mph within fifty miles of an insured property, satellite systems record the event, AI algorithms verify and analyze the data, and digital payment networks initiate instant payouts. If rainfall surpasses ten inches in twenty-four hours, these same integrated systems detect, confirm, and process the trigger conditions, enabling immediate compensation. Similarly, when an earthquake registers a magnitude of 6.0 or higher within a defined radius, satellite imaging, AI assessment, and digital payment infrastructure work in tandem to validate the event and deliver funds within hours.

This approach eliminates the documentation requirements that create the cruel paradox of traditional disaster insurance. Policyholders don't need to prove their losses with destroyed documents or wait for adjusters to assess damage that may be inaccessible for weeks. The payout triggers are based on objective, independently verifiable data that no disaster can destroy.

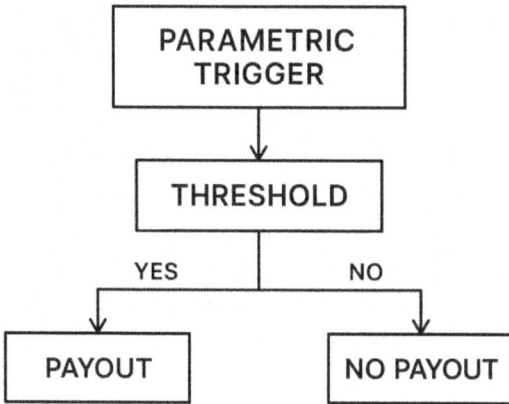

PARAMETRIC
TRIGGER

↓

THRESHOLD

YES   NO

PAYOUT   NO PAYOUT

THEORETICAL ILLUSTRATION

THEORETICAL ILLUSTRATION ONLY:

This conceptual flowchart represents how parametric trigger systems could theoretically operate. This is a speculative illustration of the decision-making process and does not describe existing automated systems currently deployed at a commercial scale.

My research into parametric insurance applications reveals significant theoretical advantages over traditional approaches. Speed represents the most obvious benefit, with theoretical response times of hours rather than months. But the advantages extend to objectivity, scalability, transparency, and efficiency.

Consider a theoretical hurricane scenario based on my research framework. As Hurricane Sarah approaches the Gulf Coast, satellite monitoring systems track its intensity and projected path. Then, AI forecasting models analyze potential impact zones and begin calculating trigger probabilities for parametric insurance policies in the region.

When Hurricane Sarah makes landfall as a Category 4 storm, weather monitoring stations record wind speeds exceeding predetermined thresholds. Parametric insurance systems automatically get activated for policies within the affected zones. Lastly, electronic notifications inform policyholders that their coverage has been triggered and that payments are being processed.

Within hours of landfall, while traditional insurance adjusters are still waiting for the storm to pass, parametric insurance funds are flowing to disaster victims. These immediate resources enable rapid response activities, securing contractors while they're available, purchasing supplies before demand drives up prices, and beginning the recovery process while infrastructure remains functional.

The theoretical benefits extend beyond individual policyholders to entire communities. Businesses with parametric coverage could reopen quickly, maintaining employment and sustaining economic activity throughout recovery. Residents with immediate insurance funds could begin repairs promptly, helping to prevent secondary damage caused by exposure and neglect.

My research suggests that communities with widespread adoption of parametric insurance could recover from disasters in weeks and months, rather than years, maintaining economic vitality and social cohesion that traditional insurance systems often fail to preserve.

Chapter 4:
Real-World Applications

The theoretical framework I have developed can be applied to various types of disasters, each with unique characteristics and specific requirements. My research has identified specific applications for hurricanes, wildfires, floods, and earthquakes, demonstrating the versatility of AI-powered parametric approaches.

Hurricane applications perhaps represent the most straightforward parametric insurance opportunity. Hurricanes are well-monitored, predictable events with clear intensity measurements and defined impact zones. Wind speed triggers could activate automatically when sustained winds exceed predetermined thresholds. Storm surge triggers could respond to measured water levels above normal. Rainfall triggers could provide coverage for flood damage based on the amount of precipitation.

Consider a theoretical scenario based on my research framework. Hurricane Michael approaches the Florida Panhandle as a rapidly intensifying storm. Satellite monitoring systems track its development, while AI models predict impact zones and intensity at landfall. Parametric insurance systems prepare for automatic activation based on wind speed measurements from weather stations along the coast.

When Hurricane Michael makes landfall with sustained winds of 155 mph, parametric triggers activate immediately for policies within fifty miles of the storm's center. Electronic notifications inform policyholders that their coverage has been triggered while the hurricane is still active. By the time the storm passes, insurance funds are already flowing to affected communities.

Wildfire applications present unique opportunities for proactive rather than reactive response. Unlike other disasters that strike suddenly, wildfires often provide warning time that could enable preventive action. Parametric wildfire insurance could activate when fires burn within predetermined distances of insured properties, providing immediate funds for fire protection measures rather than just post-disaster recovery.

My research framework envisions wildfire parametric insurance that could transform fire protection from reactive recovery to proactive defense. When satellite systems detect fire ignition, AI analysis determines threat levels for insured properties based on proximity, wind conditions, topography, and fuel loads. However, accurately predicting fire behavior and spread patterns remains challenging, even with advanced modeling, as fires can change direction rapidly in response to wind shifts and terrain features that may not be fully captured by current AI systems. Nevertheless, parametric triggers activate automatically when fires approach insured properties, providing immediate funding for fire protection services, evacuation costs, and temporary relocation.

This proactive approach could enable property owners to hire fire protection services, create defensible space, and implement protective measures while fires are still distant. The result could be tremendous: reduced property damage, fewer lives lost, and lower overall insurance costs through prevention, rather than recovery.

Flood applications face unique challenges due to the variety of flood types and causes. Flash floods develop rapidly with little to no warning, while river floods may build over days or weeks. Coastal flooding from hurricanes combines storm surge with rainfall, while urban flooding results from overwhelmed drainage systems.

My research framework addresses these challenges through multiple trigger types, tailored to different flood scenarios. Rainfall triggers could provide coverage for flash flooding based on precipitation measurements exceeding specified thresholds. Similarly, river level triggers can activate when water levels exceed the flood stage at designated monitoring stations, while storm surge triggers can respond to coastal flooding caused by hurricanes or severe storms.

The theoretical benefits of parametric flood insurance are particularly compelling, given the documentation challenges that floods create. Floodwaters destroy the records and evidence needed for traditional insurance claims, creating a cruel paradox where victims must prove damages and losses with documentation destroyed in the same floods. Thus, parametric triggers based on objective rainfall and water level measurements could eliminate these documentation requirements entirely.

Earthquake applications present both opportunities and challenges for parametric insurance. Earthquakes strike without warning, making a proactive response impossible. But they also provide clear, objective measurements through seismic monitoring networks that could enable rapid parametric response.

Magnitude triggers could activate payouts based on earthquake intensity measurements from seismic monitoring stations. Ground motion triggers could respond to peak ground acceleration levels that correlate with building damage. Duration triggers could provide coverage based on shaking duration exceeding predetermined thresholds.

The speed advantages of parametric earthquake insurance could be particularly valuable, given the time-sensitive nature of earthquake response. Immediate funding could enable rapid structural assessments, temporary housing arrangements, and emergency repairs that prevent secondary damage.

Chapter 5:
The Human Impact

Behind every insurance claim lies a human story of loss, hope, and resilience. The theoretical transformation I envision through AI-powered parametric insurance is ultimately about serving human needs more effectively than the current systems allow.

Consider the composite story of Jennifer Martinez, based on patterns I have observed in my research across multiple disaster recovery studies. Jennifer lived in Paradise, California, when the Camp Fire struck in November 2018. She had eight minutes to evacuate her home of thirty years, grabbing only her purse and her cat before fleeing the approaching flames.

The fire destroyed everything Jennifer owned, her house, her possessions, her memories, and her community. But the real ordeal began after the flames were extinguished. Her insurance company required detailed documentation of her losses, including receipts for items purchased decades earlier, photographs of possessions that had been destroyed, and proof of ownership for belongings that had been reduced to ashes.

Jennifer spent months living in a FEMA trailer while reconstructing her life from memory and bank statements.

She hired contractors who demanded payment up front but disappeared with her money. She competed with thousands of other fire victims for limited supplies and services, watching prices soar as demand overwhelmed supply. By the time her insurance claim was finally settled two years later, the cost of rebuilding had doubled, and her contractor had declared bankruptcy.

Now imagine how Jennifer's story might have unfolded with AI-powered parametric insurance based on my research framework. Firstly, satellite systems would have detected the Camp Fire within minutes of ignition. Next, AI analysis would have calculated fire threat levels for Jennifer's property based on proximity, wind conditions, and fuel loads. Then, parametric triggers would have activated automatically when the fire burned within one mile of her home.

Following this, electronic notifications would have informed Jennifer that her coverage was triggered while she was still evacuating. By the time she reached safety, insurance funds would already be flowing to her account. Instead of spending months documenting losses, she could have immediately secured temporary housing, hired reputable contractors, and purchased supplies before demand drove up prices.

The psychological benefits would have been as important as the financial ones. Rather than enduring months of uncertainty about whether her claim would be approved,

Jennifer would have known immediately that help was coming. Instead of disputing with adjusters over the value of destroyed possessions, she could have focused on rebuilding her life and community.

ECONOMIC DEVELOPMENT

COMMUNITY CAPACITY

COMMUNITY RESILIENCE

SOCIAL CAPITAL

INFORMATION AND COMMUNICATION

THEORETICAL ILLUSTRATION

THEORETICAL ILLUSTRATION ONLY:

This conceptual framework diagram represents how community resilience theoretically depends on the integration of multiple factors. This is a speculative model illustrating how various elements could interact, and it does not describe existing integrated systems or operational frameworks.

My research suggests that the community-level impacts of parametric insurance could be even more significant than the individual benefits. Communities with widespread parametric coverage could recover from disasters in months rather than years, maintaining the economic vitality and social cohesion that traditional insurance systems often fail to preserve.

Businesses with parametric coverage can reopen quickly, thereby maintaining employment and economic activity during the recovery period. Residents with immediate insurance funds could begin repairs promptly, preventing the neighborhood blight that often follows disasters. Schools and community facilities could resume operations rapidly, maintaining the social infrastructure that holds communities together.

Furthermore, the economic multiplier effects could be substantial. Every dollar of immediate insurance payout could generate multiple dollars of economic activity through rapid recovery spending. Communities that recover quickly attract new investment and residents, while those that struggle with prolonged recovery often face permanent decline.

But the most important impact would be the human experience. Disasters are traumatic enough without adding the additional trauma of bureaucratic battles and financial uncertainty. Parametric insurance could eliminate much of the secondary trauma that current systems inflict on disaster victims, allowing them to focus on healing and rebuilding, rather than fighting for the help they deserve.

Chapter 6:
Building the Future

The transformation I envision will not happen automatically. It requires deliberate action from individuals, companies, and governments to build the systems and policies that could make AI-powered parametric insurance a reality.

For insurance industry leaders, the opportunity is both significant and urgent. Climate change is increasing the frequency and intensity of disasters, while traditional insurance systems are struggling to manage the growing volume of claims. Thus, companies that develop effective parametric insurance capabilities could gain the first-mover advantage and a substantial competitive edge in resilience, efficiency, and customer satisfaction.

The technical requirements are substantial, but achievable. Insurance companies would need to develop partnerships with satellite imagery providers, AI technology companies, and weather monitoring services. They would need to create new underwriting models that price parametric risks accurately while maintaining profitability. They would also need to educate consumers about the benefits and limitations of parametric insurance.

But the potential rewards justify the investment. Parametric insurance could dramatically reduce claim processing costs while improving customer satisfaction and community outcomes. Companies that master these capabilities could expand into new markets and customer segments while building stronger relationships with existing policyholders.

For technology companies, the parametric insurance market represents a significant growth opportunity. The integration of satellite imagery, AI analysis, and parametric triggers requires sophisticated technology platforms that few companies currently provide. Early movers in this space could establish dominant positions in a rapidly growing market.

The development requirements span multiple technical domains. Satellite imagery companies need to provide real-time disaster monitoring capabilities with sufficient resolution and frequency for insurance applications. AI companies must develop machine learning algorithms that can assess disaster damage accurately and fairly under real-world conditions. However, ensuring AI fairness and accuracy across diverse communities, property types, and disaster scenarios remains a significant technical challenge that requires extensive testing and validation. Payment processing companies need to create systems capable of handling large volumes of automatic payouts during disaster events.

For policymakers and regulators, the challenge lies in enabling innovation while maintaining appropriate consumer protections. Parametric insurance represents a significant departure from traditional insurance products, requiring new regulatory frameworks that balance innovation with consumer safety.

State insurance commissioners could create regulatory sandboxes that allow controlled testing of parametric insurance products with appropriate consumer protections. Federal policymakers could establish national standards for parametric insurance while maintaining state regulatory authority. International coordination could address global climate risks through a shared monitoring infrastructure and standardized parametric insurance.

The policy framework I envision includes expedited approval processes for parametric products that meet federal standards, consumer education requirements to ensure understanding of parametric insurance benefits and limitations, quality assurance standards for AI systems used in insurance applications, and enhanced capital requirements for companies offering parametric products.

For communities and emergency managers, parametric insurance could transform disaster preparedness and response planning. Communities with widespread parametric coverage could develop more effective recovery plans that assume immediate insurance funding, rather than prolonged uncertainty.

Emergency managers could coordinate response plans with parametric insurance activation procedures, ensuring that insurance funds support rather than compete with emergency response activities. Community leaders could advocate for parametric insurance adoption while educating residents about the benefits of faster disaster recovery.

The integration opportunities are substantial. Parametric insurance systems could share data with emergency management agencies, providing real-time information about disaster impacts and recovery needs. Insurance payouts could be coordinated with emergency assistance programs, ensuring that help reaches those who need it most without duplication or gaps.

Chapter 7:
Overcoming Challenges

The path to AI-powered parametric insurance faces significant challenges that must be acknowledged and addressed. My research has identified a combination of technical, regulatory, social, and economic obstacles that could slow or even prevent the transformation I envision.

Technical challenges focus on ensuring that AI systems operate accurately and fairly under real-world conditions. Laboratory demonstrations of 90-95% accuracy in controlled environments may not translate to comparable performance amid the chaos and complexity of actual disasters. However, applying these laboratory results to real-world disaster scenarios at commercial scale remains an unsolved challenge, requiring extensive validation under diverse conditions, varying weather patterns, different disaster types, and complex damage scenarios that may not match training data. AI systems must therefore be tested extensively to ensure reliability when lives and livelihoods depend on their decisions.

Bias represents a particularly serious concern. Machine learning algorithms can perpetuate or amplify existing biases in training data, potentially leading to unfair treatment of different communities or demographic groups. Parametric insurance systems must be designed and tested to ensure equitable treatment regardless of income, race, geographic location, or other factors.

Quality control mechanisms are essential for maintaining public trust in automated systems. AI-powered parametric insurance must incorporate human oversight, audit trails, and appeals processes that enable the correction of errors or unfair decisions. The systems must be transparent enough for regulators and consumers to understand how decisions are made.

Regulatory challenges involve developing appropriate oversight for AI-powered insurance systems while enabling innovation. Current insurance regulations were designed for traditional products and may not accommodate parametric innovations effectively. Regulators must strike a balance between consumer protection and innovation support, ensuring that new products meet consumer needs without creating new risks.

The fragmented nature of US insurance regulation creates additional complexity. Parametric insurance products may require approval in all fifty states, potentially taking decades for nationwide deployment. Federal coordination mechanisms could accelerate this process while maintaining appropriate state oversight.

Consumer protection standards must be adapted for parametric insurance products. Traditional insurance regulations focus on claim handling procedures that don't apply to automatic parametric payouts. New standards must address trigger transparency, basis risk disclosure, and quality assurance for AI-powered systems.

Social challenges involve building public trust in automated disaster response systems. Many people are understandably skeptical of AI-powered decision-making, particularly for something as important as disaster insurance. Parametric insurance systems must demonstrate reliability, fairness, and transparency to gain public acceptance.

Consumer education represents a significant challenge. Parametric insurance differs fundamentally from traditional coverage, and consumers must understand both the benefits and limitations of trigger-based payouts. Basis risk, the possibility that triggers don't perfectly match individual losses, must be clearly explained and appropriately managed.

Cultural resistance to change may slow adoption even when parametric products offer clear advantages. Insurance is a conservative industry, and both companies and consumers may prefer familiar approaches even when they perform poorly. Overcoming this resistance requires demonstrating clear benefits through pilot programs and success stories.

Economic challenges involve ensuring that parametric insurance remains affordable and accessible while providing adequate coverage. At the same time, the technology infrastructure required for AI-powered systems involves significant costs that must be recovered through premiums. Therefore, pricing models must balance affordability with sustainability to ensure long-term viability.

Market concentration could become a concern if only large companies can afford the technology investments required for parametric insurance. Smaller insurers may need access to shared platforms or technology services to remain competitive. Thus, regulatory frameworks must ensure that innovation doesn't reduce competition or consumer choice.

Reinsurance markets must adapt to parametric insurance models, developing new approaches to risk sharing and capital allocation. Traditional reinsurance contracts may not accommodate parametric triggers effectively, requiring innovation in risk transfer mechanisms.

Chapter 8:
The Policy Framework

Realizing the vision of AI-powered parametric insurance requires comprehensive policy support at the federal, state, and local levels. My research has identified specific policy interventions that could accelerate development, while maintaining appropriate consumer protections.

At the federal level, leadership could provide crucial coordination and support for parametric insurance innovation. Establishing clear national standards for parametric insurance products could create consistency across state lines while maintaining state regulatory authority. Furthermore, federal research funding could accelerate the development and validation of AI-powered systems. In addition, data-sharing protocols could facilitate access to weather, satellite, and disaster information needed for effective parametric triggers

Moreover, the Federal Emergency Management Agency (FEMA) could play a key role in integrating parametric insurance with national disaster response systems. For example, FEMA data on disaster impacts could help validate parametric triggers and improve overall system accuracy. Enhanced coordination between insurance payouts and federal disaster assistance could ensure more efficient use of resources and help prevent duplication of efforts.

Additionally, the National Weather Service provides essential data for parametric insurance systems. Federal policies should ensure appropriate access to weather monitoring information while maintaining data quality and reliability. Continued federal investment in weather monitoring infrastructure could improve the accuracy and coverage of parametric triggers.

At the state level, insurance commissioners hold the primary responsibility for regulating parametric insurance products. To foster innovation, regulatory sandboxes could provide controlled environments for testing new approaches with appropriate consumer protections. Similarly, expedited approval processes could accelerate the deployment of products that meet federal standards.

Interstate coordination mechanisms could also reduce the complexity of multi-state insurance operations. Compacts between states could enable coordinated approval processes for parametric products, reducing the time and cost of regulatory compliance. Shared regulatory standards and procedures could help maintain consumer protection while supporting ongoing innovation.

From a consumer standpoint, protection enhancements must address the unique characteristics of parametric insurance. Transparency requirements could ensure clear disclosure of trigger conditions, payout calculations, and basis risk limitations. Fairness standards could prevent discriminatory pricing or coverage decisions, while appeals processes could provide recourse for consumers who believe they have been treated unfairly.

Equally important, quality assurance standards for AI systems represent a critical regulatory need. Requirements for algorithm validation, bias testing, and ongoing performance monitoring could ensure that automated systems serve consumer needs fairly and effectively. Moreover, audit trails and transparency measures could enable stronger regulatory oversight of AI-powered decisions.

At the local level, governments and emergency management agencies could integrate parametric insurance considerations into disaster preparedness and response planning. Zoning and building codes could account for parametric insurance availability and requirements. In addition, emergency response plans could coordinate with insurance payout systems to optimize resource deployment.

Community resilience planning could also incorporate parametric insurance as a tool for faster recovery and reduced disaster impacts. Local leaders could advocate for parametric insurance adoption while educating residents about the benefits of improved disaster response systems.

On the international stage, the global dimension of climate risk requires coordination beyond national borders. Global climate monitoring systems could provide data for parametric triggers worldwide. Developing international standards for parametric insurance could facilitate cross-border coverage and risk sharing. Finally, stronger coordination between developed and developing countries could extend parametric insurance benefits to vulnerable populations worldwide.

Chapter 9:
A Speculative Vision for 2030

Speculative Future Scenario Disclaimer:

This chapter presents one possible future scenario based on optimistic technological development assumptions and successful policy implementation. This represents speculative futurism, rather than guaranteed outcomes, intended to illustrate potential possibilities if current research trends continue and implementation challenges are successfully addressed.

Looking ahead to 2030, I envision a world where disaster response operates as an integrated system rather than a collection of separate, slow-moving processes. This speculative vision is grounded in current technological capabilities and policy trends, representing an achievable future if significant investments and coordination efforts succeed, rather than science fiction.

By 2030, satellite constellations could provide continuous monitoring of disaster risks and impacts across all geographic areas, assuming continued investment in space-based infrastructure and successful integration of multiple monitoring systems. Advanced AI systems could analyze multiple data streams to predict disaster impacts with unprecedented accuracy and lead time, though this assumes successful resolution of current challenges in AI reliability, bias detection, and real-world validation. Parametric insurance systems could activate automatically based on objective triggers, delivering financial assistance within hours of disasters, provided that regulatory frameworks evolve to accommodate these innovations and consumer acceptance grows sufficiently.

Integrated communication platforms could coordinate between insurance companies, emergency managers, government agencies, and affected communities. Adaptive learning mechanisms could continuously improve system accuracy and effectiveness based on real-world performance data.

Consider how hurricane response could evolve by 2030 based on my speculative research framework. Hurricane Sarah forms in the Atlantic and begins tracking toward the US coast. Satellite monitoring systems detect its formation and track its projected path with unprecedented precision. AI forecasting models analyze potential impact zones and begin calculating trigger probabilities for parametric insurance policies throughout the region.

As Sarah intensifies and approaches landfall, parametric insurance systems prepare for automatic activation based on wind speed, storm surge, and rainfall measurements. Policyholders receive notifications about potential coverage activation, allowing them to prepare for rapid response activities.

When Sarah makes landfall as a Category 4 storm, weather monitoring stations record wind speeds exceeding predetermined thresholds. Parametric insurance systems activate automatically for policies within affected zones. Electronic notifications inform policyholders that their coverage has been triggered and that payments are being processed.

Within hours of landfall, while traditional insurance adjusters would still be waiting for the storm to pass, parametric insurance funds flow to disaster victims. These immediate resources enable rapid response activities, securing contractors while they're available, purchasing supplies before demand drives up prices, and beginning recovery while infrastructure remains functional.

Detailed satellite imagery analysis provides comprehensive damage assessment within 24 hours of the storm's passage. AI systems identify properties requiring immediate attention and calculate appropriate additional payouts for severe damage. Quality control mechanisms ensure accuracy while human oversight handles complex cases.

Communities with widespread parametric coverage begin recovery immediately. Businesses reopen within days rather than months, maintaining employment and economic activity. Residents begin repairs promptly, preventing secondary damage from exposure and neglect. Schools and community facilities resume operations rapidly, maintaining social infrastructure.

By 2030, this integrated approach could be extended beyond hurricanes to encompass all major disaster types, assuming successful technology development and effective policy implementation. Wildfire threats could be detected within minutes of ignition, with parametric insurance providing immediate funding for fire protection measures. Flood impacts could be assessed through satellite imagery and sensor networks, with payouts triggered by objective rainfall and water level measurements. Earthquake response could begin within hours of seismic events, with parametric funding enabling immediate structural assessments and emergency repairs.

The economic benefits could also be transformative, though this depends on widespread adoption and successful system integration. Communities that recover quickly from disasters attract investment and population growth, while those that struggle with prolonged recovery often face permanent decline. The speed of insurance response directly correlates with long-term community resilience and prosperity.

But the most important transformation would be human. Disasters would remain traumatic, but the secondary trauma of bureaucratic battles and financial uncertainty could be largely eliminated. Disaster victims could focus on healing and rebuilding rather than fighting for the help they deserve and paid for in advance.

This speculative vision requires continued investment in technology development, policy innovation, and system integration. However, it represents an achievable future based on capabilities that exist today, waiting to be combined into something greater than the sum of their parts, provided that current research and development efforts succeed and implementation challenges are overcome.

Chapter 10:
Taking Action

The transformation I envision will not happen automatically. It requires deliberate action from individuals across multiple sectors who recognize the potential for change and commit to making it a reality.

For insurance industry professionals, the opportunity is both immediate and strategic. Companies that develop parametric insurance capabilities now could gain substantial competitive advantages as climate change increases the frequency and severity of disasters. The technical requirements are substantial but achievable through partnerships with technology companies and gradual capability development.

Insurance executives should begin by evaluating current disaster response capabilities and identifying opportunities for parametric insurance applications. Pilot programs in limited geographic areas could demonstrate effectiveness while building internal expertise. Partnerships with satellite imagery providers, AI companies, and weather monitoring services could provide the necessary technical capabilities.

Consumer education represents a critical success factor. Parametric insurance differs fundamentally from traditional coverage, and consumers must understand both benefits and limitations. Investments in clear and transparent communication about trigger conditions, payout calculations, and basis risk can build trust and adoption.

For technology companies, the parametric insurance market represents significant growth potential. The integration of satellite imagery, AI analysis, and parametric triggers requires sophisticated platforms that few companies currently provide. Early investment in these capabilities could establish dominant market positions.

Satellite imagery companies should focus on providing real-time disaster monitoring with sufficient resolution and frequency for insurance applications. AI companies should develop machine learning algorithms that can accurately and fairly assess disaster damage under diverse real-world conditions. However, ensuring AI systems work reliably across different disaster types, geographic regions, and community characteristics remains a significant technical challenge requiring extensive validation and testing. Furthermore, payment processing companies should create systems capable of handling large volumes of automatic payouts during disaster events.

Quality assurance and bias detection represent critical technical requirements. AI systems used for insurance decisions must demonstrate fairness across different communities and demographic groups. Transparency and explainability features could enable regulatory approval and consumer trust.

For policymakers and regulators, the challenge involves enabling innovation while maintaining consumer protection. Regulatory sandboxes could provide controlled environments for testing parametric insurance products, with appropriate safeguards in place. Expedited approval processes could accelerate deployment while maintaining oversight.

State insurance commissioners should engage with insurance companies developing parametric products by participating in industry conferences and working groups focused on innovation. Interstate coordination mechanisms could reduce regulatory complexity while maintaining consumer protection standards.

Federal policymakers could establish national standards for parametric insurance while preserving state regulatory authority.

Data-sharing frameworks could facilitate access to weather and disaster information needed for effective parametric triggers. Research funding could accelerate the development and validation of AI-powered systems.

For community leaders and emergency managers, parametric insurance could transform disaster preparedness and response planning. Communities with widespread parametric coverage could develop more effective recovery plans that assume immediate insurance funding rather than prolonged uncertainty.

Emergency managers should coordinate response plans with parametric insurance activation procedures, ensuring that insurance funds support rather than compete with emergency response activities.

Community education programs could inform residents about the benefits of parametric insurance while building support for its adoption.

Advocacy with state and federal policymakers could advance policy frameworks that enable parametric insurance innovation. Participation in industry conferences and working groups could help share community perspectives with system developers.

For academic researchers, parametric insurance represents a rich field for investigation across multiple disciplines. Technology validation studies could assess AI accuracy and bias in disaster assessment applications.

Policy analysis could evaluate regulatory frameworks that enable or hinder innovation. Social impact research could examine community-level effects of parametric insurance adoption.

Collaboration with industry partners could provide access to real-world data and implementation challenges. International comparative studies could identify best practices and policy approaches from different markets.

Interdisciplinary research could address the complex interactions between technology, policy, and social systems.

For individual consumers, the most important action involves education and advocacy. Understanding parametric insurance benefits and limitations could inform purchasing decisions as products become available. Advocacy with insurance companies and policymakers could accelerate development and deployment.

Community engagement can help build support for the adoption of parametric insurance while ensuring that products effectively meet local needs.

Participation in pilot programs could provide valuable feedback for system improvement.

The timeline for transformation depends on the level of commitment and coordination across these different sectors. With strong leadership and investment, significant progress could be achieved within five years, while comprehensive implementation could be possible within a decade.

But the transformation requires starting now. Climate change is increasing the frequency and severity of disasters, while traditional insurance systems are becoming less capable of handling the resulting challenges. The necessary technologies for parametric insurance are available today, waiting to be integrated into comprehensive solutions.

The question is not whether transformation is possible; the capabilities exist. The question is whether we will choose to build systems that serve human needs more effectively than current approaches, which hinder the insurance process, take months to settle claims, and years to reimburse policyholders. That choice is before us now, and the decisions we make today will determine the future of disaster response for generations to come.

Conclusion:
When Seconds Count

As I complete this research framework, I return to the question that started my journey: *"The flood happened in one day. Why does the insurance take one year?"* That Nigerian farmer's simple question exposed a fundamental flaw in how we approach disaster response, launching my research into how emerging technologies could enable a dramatically better approach.

Through eight years of work across Nigerian and US insurance markets, I have seen both the failures of current systems and the potential for transformation. The vision I present in this book is not inevitable; it requires choice, investment, and commitment from individuals, companies, and governments. However, it has become achievable based on the technologies and principles that exist today.

The framework I have developed integrates existing capabilities rather than requiring breakthrough innovations. To begin with, satellite monitoring systems provide unprecedented Earth observation capabilities. Furthermore, artificial intelligence can analyze vast amounts of data with remarkable speed and accuracy in controlled environments, though translating these capabilities to real-world disaster scenarios at commercial scale remains an ongoing challenge requiring continued research and validation. In addition, parametric insurance principles have proven effective in limited applications. Finally, digital payment systems can transfer funds globally within minutes.

The innovation lies not in creating new technologies, but in combining existing capabilities into an efficient and effective comprehensive disaster response system that could deliver help at the speed of crisis rather than the pace of bureaucracy.

The potential benefits of this system extend far beyond individual policyholders to entire communities and the economic system. Faster disaster recovery could maintain employment, preserve social infrastructure, and attract investment that builds long-term resilience. The economic multiplier effects of immediate insurance response could generate substantial returns on the investments required for system development.

The most significant benefits would center on the human experience. Disasters are already deeply traumatic, and the added stress of bureaucracy and financial uncertainty only intensifies that pain. Thus, AI-powered parametric insurance can help remove much of this unnecessary hardship, enabling victims to focus on recovery and rebuilding their lives instead of battling for the assistance they need and paid for in advance.

The challenges are real and significant. Technical systems must work accurately and fairly under real-world conditions. Regulatory frameworks must balance innovation with consumer protection. Social acceptance must be built through demonstration of reliability and fairness. Economic models must ensure affordability and accessibility.

However, these challenges are ultimately surmountable with the right investment and commitment. Together, the policy framework I have outlined provides a roadmap for addressing regulatory requirements while enabling innovation. Likewise, the technical approaches I have described build on proven capabilities while acknowledging current limitations. Finally, the implementation strategy I have proposed offers a gradual path from pilot programs to comprehensive deployment.

The transformation I envision represents more than technological change; it represents a fundamental shift in how we think about disaster response. Instead of accepting that disasters must be followed by prolonged suffering, we could choose to build systems that enable immediate recovery. Instead of treating disaster insurance as a necessary evil, we could make it a tool for community resilience and economic strength.

This shift requires leadership from multiple sectors working together toward common goals. Insurance companies must invest in new capabilities while maintaining financial stability. Technology companies must develop reliable, fair systems that serve human needs. Policymakers must create frameworks that enable innovation while protecting consumers. Communities must engage with new approaches while maintaining appropriate skepticism.

The research framework I have presented provides a foundation for this collaboration, but it is only the beginning. The specific technologies and implementation details will undoubtedly evolve as research continues and real-world experience accumulates. The goal is not to implement my particular vision, but to create systems that serve human needs more effectively than current approaches.

My contribution to this effort draws on unique experience across Nigerian and US insurance markets, providing insights into both the universal challenges of disaster insurance and the varying approaches to addressing them. This cross-market perspective reveals that the problems with traditional disaster response are not unique to any single system or level of development.

The solutions I propose can be adapted for various markets and contexts, ranging from advanced economies with sophisticated infrastructure to emerging markets with limited traditional insurance capacity. The fundamental principles, such as speed, objectivity, scalability, and transparency, apply universally, even as specific implementations vary based on local conditions and capabilities.

As I look toward the future, I see both urgency and opportunity. Climate change is increasing the frequency and severity of disasters while traditional insurance systems become less capable of handling the resulting challenges. The gap between the speed of crisis and the pace of response continues to widen, creating more suffering and economic damage with each passing year.

But the technologies needed for transformation exist today, and the policy frameworks for enabling change are within reach. The only question is whether we will choose to build something better than what we have now.

The farmer in Kebbi State who asked why insurance takes a year when floods happen in a day deserves a better answer than the bureaucratic explanations that traditional systems provide. Hurricane victims waiting in FEMA trailers deserve faster help than the current processes deliver. Wildfire survivors rebuilding their lives deserve immediate support rather than prolonged uncertainty.

They all deserve insurance that works at the speed of crisis rather than the pace of bureaucracy. The technology exists to provide it. The knowledge exists to implement it. The policy frameworks exist to enable it.

The only question is, do we really want to make it a reality?

When seconds count, we cannot afford to wait. The future of disaster response is being written today through the research, policy, and investment decisions we make now. I hope this framework contributes to writing a better future for the millions of people who will face disasters in the years to come.

The transformation is possible. The question is whether we will make it happen.

About the Author

Bunmi Ogunwusi brings a unique perspective to disaster insurance research through her experience across Nigerian and US insurance markets. As a researcher and consultant with over eight years of experience in analyzing insurance systems, she has developed innovative frameworks for understanding and improving disaster response mechanisms.

Her research focuses on the intersection of artificial intelligence, parametric insurance, and disaster recovery, with particular emphasis on bridging insights from emerging and developed insurance markets. Through her work as a licensed insurance professional and business analyst, she has identified critical gaps in current disaster response systems and developed theoretical frameworks for addressing these challenges.

This book represents the culmination of her research into how AI-powered parametric insurance could transform disaster recovery, drawing on her unique experience in both Nigerian agricultural insurance consulting and US commercial insurance operations. Her cross-market perspective provides insights into both the universal challenges of disaster insurance and the innovative potential of emerging technologies.

Bunmi holds a bachelor's degree in Banking and Finance and has earned professional certifications in business analysis and insurance. She has worked as a licensed insurance professional in the United States and previously provided consulting services to insurance companies in Nigeria, focusing on agricultural risk assessment and product development.

Her current research extends her practical experience into academic frameworks for insurance innovation, with particular focus on parametric insurance applications and AI-powered disaster response systems. She is developing academic publications on these topics while continuing her professional work in insurance and business analysis.

Research Disclaimers:

This book presents theoretical research frameworks and applications based on current technological capabilities and documented disaster patterns. All scenarios and case studies are research simulations designed to illustrate potential applications rather than descriptions of existing systems. The integration of AI, satellite monitoring, and parametric insurance into comprehensive disaster response systems remains largely theoretical and requires continued research, development, and deployment. Furthermore, implementation would require extensive regulatory approval and consumer protection measures not currently in place. All diagrams and illustrations are theoretical representations of conceptual frameworks rather than descriptions of existing commercial systems.

www.ingramcontent.com/pod-product-compliance
Lightning Source LLC
Chambersburg PA
CBHW071516210326
41597CB00018B/2772